Answer Key to accompany
Chemistry: A Graphing Calculator Approach

Page 3
1. zinc fits post 1982 pennies, copper fits pre 1982 pennies
2. to make sure the mass is constant
3. No, some data could be incorrect because of inaccurate reading, meniscus may not be read when recording the volume, balance may not be zeroed out before reading mass etc...
4. Pre 1982, 60 pennies, 186 grams
   Post 1982, 60 pennies, 150 grams
5. Light and heavy are not good terms to use since they only indicate weight or mass. Dense is more descriptive since it describes a ratio of mass and volume. Density is an intrinsic physical property which is not dependent on the size of the substance.

Page 7

32.14g	32.16g	1.	Trial 1
142.76g	138.24g		ΔT(water) = 25.54 - 23 = 2.54
110.62g	106.08g		ΔT(metal) = 99.02 - 25.54 = 73.48
23°C	23.2°C		
99.02°C	99.24°C		Trial 2
25.54°C	25.47°C		ΔT(water) = 25.47 - 23.2 = 2.27
100g	100g		ΔT(metal) = 99.24 - 25.47 = 73.77

Page 8
2. Trial 1 Heat gained (water) = 4.18 * 2.54 * 100 = 1061.72
   Trial 2 Heat gained (water) = 4.18 * 2.27 * 100 = 948.86
3. Trial 1 1061.72 = sh * 73.48 * 110.62
         .1306 = sh
   Trial 2 948.86 = sh * 73.77 * 106.08
         .1213 = sh
4. Avg = (.1306 + .1213)/2 = .1260

Page 9
1. .899, .444, .390, .385, .129
3. %error = ((experimental - actual)/actual)*100
         = ((.1260 - .129)/.129)*100 = 2.33%

4. It was assumed that there was no heat loss from the foam cup. It was also assumed that the metal did not lose any heat as it was poured from the cup.

5. Yes, it is a value that is constant for the same substance under different conditions and that varies for different substances under identical conditions. Specific heat then is a physical property.

Page 10
graph c

Page 11
3. 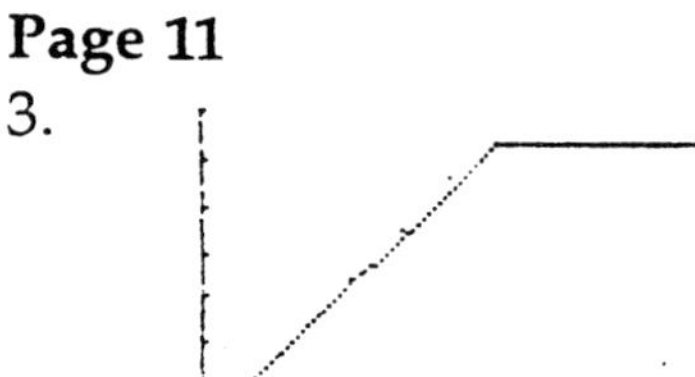   4. 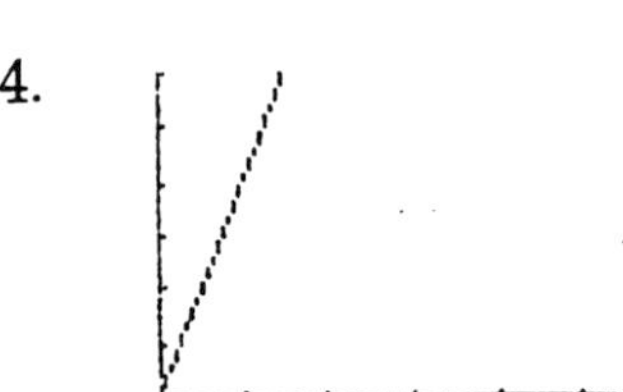

4. 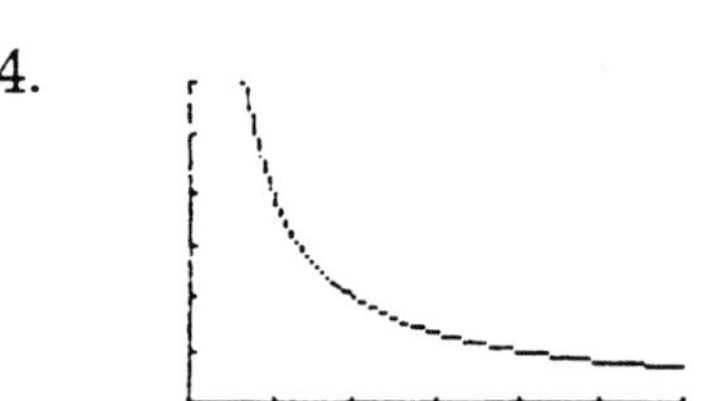

Page 15

1.1 4.8	274	.018	
23.2	5.2	296	.018
40.1	5.4	313	.017
61.4	5.8	334	.017
79	6.2	352	.018
99.9	6.5	373	.017

Page 16
1. Yes, within the limits of experimental error, the ratio of volume to temperature is a constant as Charles law would predict.
2. A change in pressure would cause a change in volume. This would interfere with determining the temperature-volume relationship.
3. a. inability to measure the height of the air column accurately.
   b. expansion of the glass tube, giving a smaller than expected volume.
   c. not allowing enough time for the gas to achieve a temperature equilibrium with the water bath.

Page 19

0	0	0
1	150	150
2	75	150
3	50	150
4	37.5	150
5	30	150
6	25	150
7	21.4	150
8	18.7	150
9	16.7	150
10	15	150

Page 20

2. As the pressure increases the volume decreases. This is an inverse relationship.
3. It is the same for each one.
4. The gas particles would get close enough together to become a liquid or even a solid.

Page 24

375	63.9
400	48.2
405	46.8
415	46.6
425	53.1
440	59.9
455	71.6
470	80.1
490	82.6
500	82.9
520	72.2
530	67.4
540	62.8
550	58.4
570	54.1
575	53.2
580	52.9
600	55.9
625	66.2

Page 25

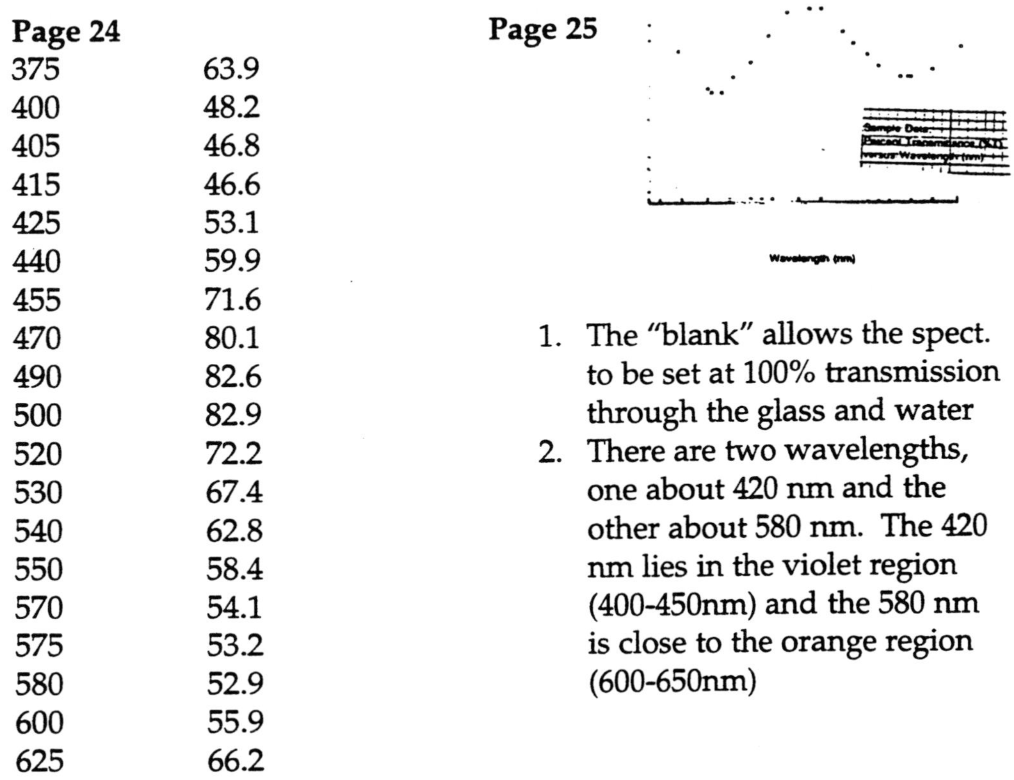

1. The "blank" allows the spect. to be set at 100% transmission through the glass and water
2. There are two wavelengths, one about 420 nm and the other about 580 nm. The 420 nm lies in the violet region (400-450nm) and the 580 nm is close to the orange region (600-650nm)

3. Colored solutions absorb light of a complementary color. The compliment of blue is orange. The blue $Cr(NO_3)_3$ solution absorbs orange light. A red solution would absorb green light and transmit red.

4. Experimental error = $\dfrac{\text{experimental - actual}}{\text{actual}} * 100$

Page 37

1. least   most   most
   3rd   least   3rd
   2nd   least   least
   most   most   2nd
2. In some chemicals solubility will greatly increase. Some chemicals show hardly any change.
3. The granulated salt will dissolve much faster.
4. Stirring greatly accelerated the solution process.
5. Heat, stirring, particle size and chemical type
Yes, the chemical will be affected by one of these factors. No, it does not work the same for all chemicals. Some chemicals are soluble only in certain substances.

Page 41

1. a. lithium fluoride will resemble NaCl
   b. $BeCl_2$ will resemble $CaCl_2$
   c. KBr will resemble NaCl
   d. MgI_2 will resemble $CaCl_2$
2. It did not disassociate. Since it did not disassociate it only added one particle per molecule not two or three like the other chemicals we used.
3. There would be more elevation up to a point. Once that point is reached the BP will not be raised any higher.

Page 44

   a. 2M   b. .4 M   c. 1M
1. a. 2 m   b. .4 m   c. 1m

Page 52

1. q = 5580 * 980 * .448 = 2449.8 Kj
4. q = 10 * 90 * 4.18 = 3.762 Kj
   q = 10 * 2260 = 22.Kj
   q = 10 * 50 * 2.20 = <u>1.01 Kj</u>
                   28.372 Kj
5. q = 42.5 * 2283 * 1.05 = 102 Kj
   q = 42.5 * (1 mol/27g) (291Kj/mol) = 458 Kj
   q = 42.5 * 1807 * .869 = 66.7 Kj
   q = 42.5 * (1 mol/27g)(10.5Kj/mol) = 16.5 Kj
   q = 42.5 * 635 * .9 = <u>24.3 Kj</u>
                 668 Kj

 Pencil Point Press, Inc
Fairfield, NJ 07004
1-800-356-1299
ISBN: 1-881641-33-3

 ANSWER KEY TO ACCOMPANY CHEMISTRY: A GRAPHING CALCULATOR APPROACH

Chemistry:
A Graphing Calculator Approach

David P. Lawrence
Southwestern Oklahoma State University

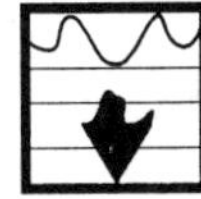

Pencil Point Press, Inc.

In memory of Wayne T. Smith and V.A. Lawrence, my Grandfathers.

This publication was developed, in part, by support of a Dwight D. Eisenhower Mathematics and Science Education Program Grant, U. S. Department of Education, Title II, P.L. 100-297, in cooperation with the Oklahoma State Regents for Higher Education. However, any opinions, findings, conclusions, or recommendations expressed herein are those of the authors, and do not necessarily reflect the views of the U.S. Department of Education or the Oklahoma State Regents for Higher Education.

Acknowledgements
I would like to thank the many Oklahoma teachers who participated in the development of this book. The participants are: Andy Evans, Tuttle High School; Keith C. Jackson, Elmore City/Pernell High School; Tom Jackson, Maud High School; Sherrie Kenney, Grant High School; Marie Pool, Clinton High School; and Candace Thomas, Watonga High School.

I am grateful to Mr. Gene Garone of Pencil Point Press, Inc. for publishing the book. I am especially thankful to Lisa B. Friesen for her help with the project. Special appreciation is extended to Southwestern Oklahoma State University and its Mathematics Department for their support during this endeavor.

About the Author
David P. Lawrence received his Ph.D. in mathematics education from the University of Oklahoma in 1989. He is privileged to have mathematics teaching experience at many levels, including high school, junior college, private college, and major university. Dr. Lawrence conducts graphing calculator workshops across the nation.

The activities provided in this book are for classroom use only.

Pencil Point Press, Inc.

277 Fairfield Road, Fairfield, New Jersey 07004

1-800-356-1299

ISBN 1-881641-24-4

HOP 10 9 8 7 6 5 4 3 2

About this book...

This workbook was first funded by a Dwight D. Eisenhower Mathematics and Science Education Grant and Southwestern Oklahoma State University, specifically for Oklahoma educators. Now presented as a national edition, the objective of this publication is to provide all chemistry teachers with materials for enhancement of chemistry instruction using graphing calculator technology or, in some cases, computer graphing software. "Turning students on" to chemistry is the main focus.

This publication is unique in at least two respects. First, it is designed to supplement the chemistry curriculum with quality student activities using any model graphing calculator as an investigative and problem-solving tool. To our knowledge, no other commercially-available publication is designed to this end. Second, the publication presents comparatively short, easy-to-use activities that promote highly-effective responses from the students.

The book is designed to be used as either as a Teacher's Resource, a student manual—or both. An answer key to accompany this book is also available to educators upon request. I hope you find the materials useful as part of your chemistry curriculum.

David P. Lawrence

Contents

An answer key is available to educators upon request.

Answer Key to Accompany
Chemistry: A Graphing Calculator Approach   ISBN: 1-881641-33-3

Other graphing calculator books available from the publisher:

Graphing Calculators: Quick and Easy The TI-81
Graphing Calculators: Quick and Easy The TI-82
Graphing Calculators: Quick and Easy The TI-85
Graphing Calculators: Quick and Easy The Sharp EL 9200/9300
Graphing Calculators: Quick and Easy The Casio fx-7000
Graphing Calculators: Quick and Easy The Casio fx-7700/8700
Graphing Calculators: Quick and Easy The Casio fx-7700GE
Graphing Calculators: Quick and Easy The Casio fx-9700GE
Graphing Calculators: Quick and Easy The HP 48G/48GX

Physics: A Graphing Calculator Approach
Biology: A Graphing Calculator Approach

The Graphing Calculator Connection: Modeling and Applications
The Graphing Calculator Connection: Investigations and Problem Solving

Applying Pre-Algebra/Algebra Using the Sharp EL 9200/9300
Applying Statistics Using the Sharp EL 9200/9300
Applying Trigonometry Using the Sharp EL 9200/9300
Applying Pre-Calculus/Calculus Using the Sharp EL 9200/9300

Skill and Practice Masters in Algebra Using the TI-82
Skill and Practice Masters in Calculus Using the TI-82
Skill and Practice Masters in Statistics Using the TI-82

*To find out more about our books or to receive a catalog of
other related materials, call **1-800-356-1299**.*

Summary of Mathematics Skills for Chemistry: A Graphing Calculator Approach

Density
- Using a table
- Measurement
- Finding an average
- Creating a statistical data set
- Drawing a scatter plot
- Graphing the best-fitting line
- Extrapolation

Specific Heat of Metal
- Measurement
- Using a formula
- Finding an average
- Creating a statistical data set
- Drawing a scatter plot

Curve Fitting
- Visualizing a graph from raw data
- Creating a statistical data set
- Drawing a scatter plot
- Working with proportional relationships
- Sketching a graph

Kinetic Theory - Charles' Law
- Measurement
- Using a table
- Using a formula
- Creating a statistical data set
- Drawing a scatter plot
- Graphing the best-fitting line

Boyle's Law

- Measurement
- Using a table
- Using a formula
- Creating a statistical data set
- Drawing a scatter plot
- Graphing the best-fitting line
- Interpolation
- Determining slope of a line

Introduction to the Spectrophotometer

- Measurement
- Using a table
- Creating a statistical data set
- Drawing a scatter plot

Freezing Point Depression Analysis of the Ice Cream Process

- Measurement
- Using a table
- Creating a statistical data set
- Drawing a scatter plot
- Determining the proportional relationship

Solubility

- Measurement
- Using a table
- Creating a statistical data set
- Drawing a scatter plot
- Graphing the best-fitting line

Boiling Point Elevation

- Measurement
- Using a table
- Creating a statistical data set
- Graphing a line

The Dreaded M's

- Using a formula
- Creating a statistical data set
- Drawing a scatter plot

Reaction Rate

- Measurement
- Using a table
- Creating a statistical data set
- Drawing a scatter plot
- Graphing the best-fitting line
- Interpolation

Heat of Fusion and Vaporization

- Using a formula
- Graphing an equation

Density

Laboratory Activity

Content: Direct linear relationships, measurement of mass and volume for calculation of density, and graphing a scatter plot.

We commonly say that lead sinks in water because "lead is heavy." We also say that cork floats on water because "cork is light." In order to determine a difference between "light" and "heavy", we must compare a ratio of mass to volume of a substance. By using mass and volume, a calculated quantity known as density may be obtained.

Are the densities of pennies minted before 1982 different from those minted after 1982? What can density tell you about the composition of the pennies?

Materials for each Lab Group:

PRE–1982 pennies (number of pennies
    should vary with each lab group)
POST–1982 pennies (number should be
    the same as pre-1982 pennies)
50 mL graduated cylinder
balance
water
graphing calculator

Procedure:

1. Make a data table to record class mass, volume, and density.

2. Using a balance, determine the mass of your PRE–1982 pennies. Repeat your measurement at least two more times. Average to determine constant mass.

3. Repeat step 2 with POST–1982 pennies.

4. Pour 25 mL water into the 50 mL graduated cylinder. Add the PRE–1992 pennies so they are covered with water. Using water displacement determine the volume of the pennies.Repeat volume measurement two more times and average results.

5. Repeat step 4 using POST–1982 pennies.

6. Using the average volume and the average mass of the two sets of pennies, determine the average density of each set. Record your results on the board for other lab groups to copy.

7. Clean and return all lab equipment and supplies to their proper location.

Graphing:

1. Set the range on your graphing calculator to Xmin: -5, Xmax = 60, Xscl = 5, Ymin = -5, Ymax = 30, Yscl: 5.

2. Create a statistical data set on your graphing calculator with the ordered pairs of class data for PRE-1982 pennies mass and volume.

3. Use your graphing calculator to produce a scatter plot for the data set formed in #2.

4. Use your graphing calculator to find the best-fitting line for the data (linear regression).

5. Repeat steps 2, 3, and 4 for the POST–1982 pennies.

1. Compare the average experimental densities of the two sets of pennies with a density table. What density values fit the experimental data?

2. Why does the procedure include three trials instead of one trial for each set of pennies?

3. Does all the data seem to contain accurate and correct measurements? What might be some causes of error?

4. Use your graphing calculator to graph the best-fitting lines found above. Extrapolate the mass for 60 pennies for each set.

5. Discuss which terms would be most appropriate to describe the two sets of pennies. LIGHT HEAVY DENSE

Specific Heat of a Metal

Laboratory Activity

Content: Scatter plots

How is heat measured? Let's use water as an example. If the water is cooled it loses heat to its surroundings. As a result, its temperature decreases. If the water is heated, its temperature increases. The quantity of heat lost or gained by water or any substance can be found from the temperature change of the substance. The heat lost or gained can be measured in joules or calories.

During a temperature change, three things determine the quantity of heat lost or gained:

1. Quantity or mass of the substance.

2. Kind of matter (element or compound).

3. Amount of temperature change.

One of the physical properties of matter is called specific heat. It is the amount of heat required to raise the temperature of one gram of the substance one Celsius degree ($1C^\circ$). The value of specific heat can be given in derived units of joules per gram per Celsius degree [$J/(g \times C^\circ)$] or calories per gram per Celsius degree [$cal/(g \times C^\circ)$].

The equation is:

Specific heat = (heat lost or gained) $\div$ (mass $\times$ Δtemperature)

The abbreviated form is:

$$c = q \div (m \times \Delta T)$$

In this experiment, groups will determine the specific heat of various metals. A heated metal sample will be added to a crude calorimeter, consisting of cool water contained in a plastic foam cup. Shortly after mixing, the water and the metal will have come to the same temperature. The foam cup acts as an insulator so that heat cannot easily escape from the calorimeter. Heat lost by the metal can be said, for the purpose of this experiment, to be equal to the heat gained by the water.

The amount of heat energy gained by the water will be calculated using the following equations:

$$\text{heat gained}_{water} = \text{specific heat}_{water} \times \text{mass}_{water} \times \Delta \text{ temperature}_{water}$$

The amount of heat lost by the metal will be calculated by a similar equation:

$$\text{heat lost}_{metal} = \text{specific heat}_{metal} \times \text{mass}_{metal} \times \Delta \text{temperature}_{metal}$$

Because heat lost should equal heat gained, a third equation can be written:

$$\text{specific heat}_{metal} \times \text{mass}_{metal} \times \Delta \text{temperature}_{metal} =$$
$$\text{specific heat}_{water} \times \text{mass}_{water} \times \Delta \text{temperature}_{water}$$

Because the specific heat of water is 4.18 Joules/g.C°, the specific heat of different metals may be calculated using the third equation.

Materials for each Lab Group:

safety goggles	1 ring stand
1 50 mL beaker	1 ring support
1 250 mL beaker	1 wire gauze
1 400 mL beaker	1 gas burner
1 100 mL graduated cylinder	centigram balance
1 large test tube	plastic foam cup
1 glass stirring rod	thermometer
1 utility clamp	water
string	

and one of the following metals in cylinder form:
lead   copper   aluminum   iron   zinc (Circle metal used)

Procedure:

1. Add 250 mL of tap water to a 400 mL beaker. Place the beaker on
    a wire gauze on a ring support clamped to a ring stand. Use a gas
    burner to bring the water to a slow boil. While the water is heating,
    proceed to step 2.

2. Determine the mass of a clean, dry 50 mL beaker to the nearest 0.01 g.
    Add the cylinder of metal to the beaker, and measure the combined
    mass of the beaker and metal cylinder to the nearest 0.01 g. Record
    these masses in data table.

3. Tie one end of the string to the utility clamp and the other end to the
    metal cylinder. Suspend the metal in the boiling water in the beaker.
    Position the metal cylinder so that it is below the level of water in the
    beaker. Allow the metal cylinder to remain in the boiling water for at
    least 10 minutes. While the metal is heating continue with next step.

4. Carefully measure out 100 mL of distilled water in a graduated cylinder,
    and pour the water into a plastic cup. Place the cup in a 250 mL beaker
    for support.

5. After the metal has been heating for at least 10 minutes, record the
    temperature of both the water in the plastic foam cup and the water in
    the boiling bath. (Assume that the temperature of the metal is the
    same as the boiling water.)

6. Remove the metal, using the clamp/string as a holder. Carefully, but
    quickly, put the heated metal cylinder into the plastic foam cup. Place
    the thermometer and a stirring rod in the cup. Use the stirring rod to
    gently move the metal in the cup. Do not use the thermometer to
    move the metal.

    Note the temperature frequently. As the temperature begins to change
    more slowly, watch the thermometer continuously so as not to miss the
    maximum temperature reached. Record this maximum temperature.

7. Pour the water off the metal and return the metal to your teacher to be
    dried.

8. If time permits, repeat the experiment.

9. Clean all lab equipment and return to its proper location.

Data:

Trial #1   Trial #2

mass of 50 mL beaker

mass of 50 mL beaker and metal cylinder

mass of metal cylinder

initial temperature of water in cup

initial temperature of metal cylinder
    (temperature of boiling water)

maximum temperature of metal cylinder
    and water

mass of water

Calculations:

1. Calculate the changes in temperature of the water (ΔT) and of the metal cylinder (ΔT) for each trial.

2. Calculate the heat gained by the water in each trial. Express heat gained in joules.

3. Because heat gained by the water is equal to heat lost by the metal cylinder, calculate the specific heat of the metal for each trial. Express specific heat in $J \div (g \times C^{\circ})$.

4. Calculate your average value for the specific heat of your metal.

1. Complete the following data from other lab students:

#	Metal used	Specific heat in J/g*C°
1	aluminum	_______________
2	iron	_______________
3	zinc	_______________
4	copper	_______________
5	lead	_______________

2. Create a two-variable data set within your graphing calculator for the ordered pairs (#, specific heat). Draw the scatter plot for the data within the range Xmin= 0, Xmax = 6, Xscl = 1, Ymin = 0, Ymax = 1, Yscl = .1.

3. Using a scatter plot obtained from your teacher, calculate the percent error in the specific heat value for each metal.

4. You assumed that the initial temperature of the metal cylinder was the same as that of the boiling water. This assumption may not have been correct, and may account for some experimental error. What other assumptions were made that may be sources of error?

5. Compare your specific heat scatter plot with the one provided by your teacher. Can specific heat be used to identify a substance?

Curve Fitting

Pre-Laboratory Activity

Content: Matching graphs from data or formulas, using scatter plots and sketching graphs.

In a chemical reaction, substances X and Y react to form substance Z. Masses of the reactants X and Y and the product Z are carefully measured. The amount of Z formed is generally not the sum of the masses of X and Y. It is assumed that for each reaction either X or Y is in excess.

EXPERIMENT 1

X	Y	Z
2	20	5.04
4	20	10.09
6	20	15.13
8	20	20.17
10	20	25.22
12	20	30.26
14	20	33.14
16	20	33.16
18	20	33.15
20	20	33.14

EXPERIMENT 2

X	Y	Z
1.4	2	2.25
1.4	4	4.50
1.4	6	6.75
1.4	8	9.00
1.4	10	11.25
1.4	12	12.60
1.4	14	12.59
1.4	16	12.61
1.4	18	12.60
1.4	20	12.61

1. Which of the following graphs would represent the results of the mass of X vs. Z for Experiment 1.

a.

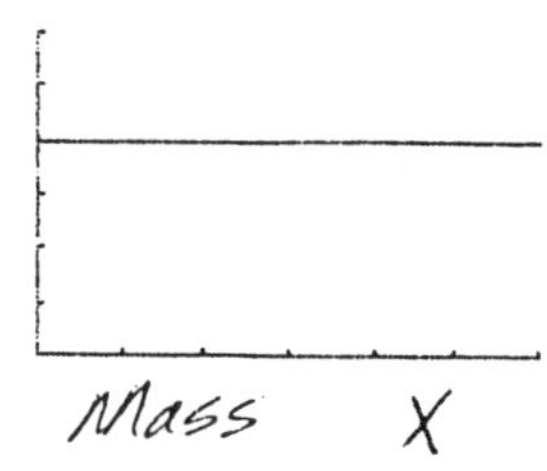

b.

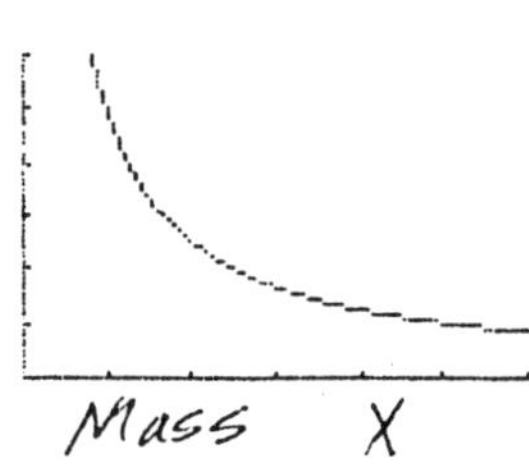

c.

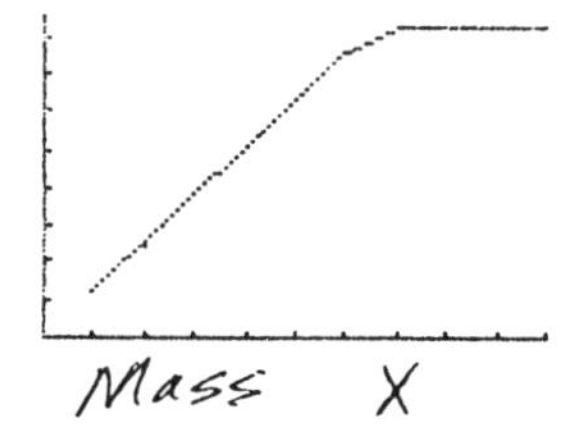

d.

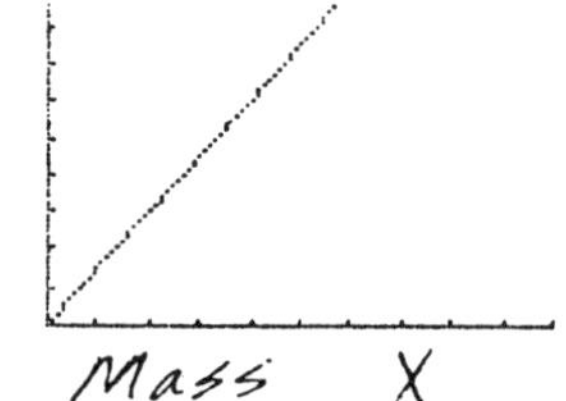

2. Place the ordered pairs (X, Z) from Experiment 1 into a statistical data set within your graphing calculator. Draw the scatter plot for the data to verify your answer in #1.

3. Place the ordered pairs (Y, Z) from Experiment 2 into a statistical data set within your graphing calculator. Draw the scatter plot for the data. Sketch the graph that represents the data on the axes provided below.

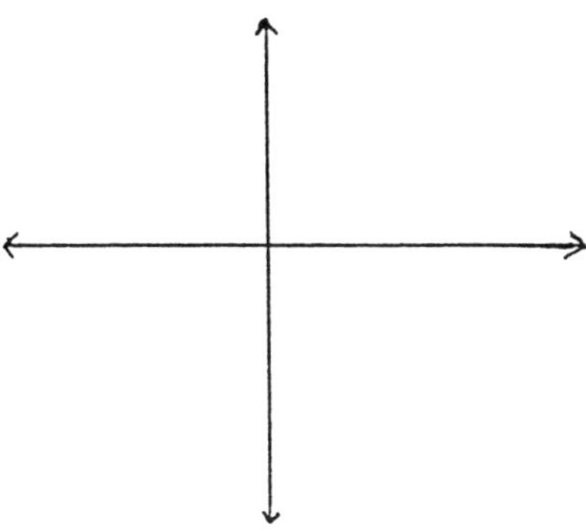

4. In working with proportional relationships you will end up with different types of graphs. Use your graphing calculator to help you sketch the graph of each relationship.

y=kx(direct proportion)   y=k/x(inverse proportion)

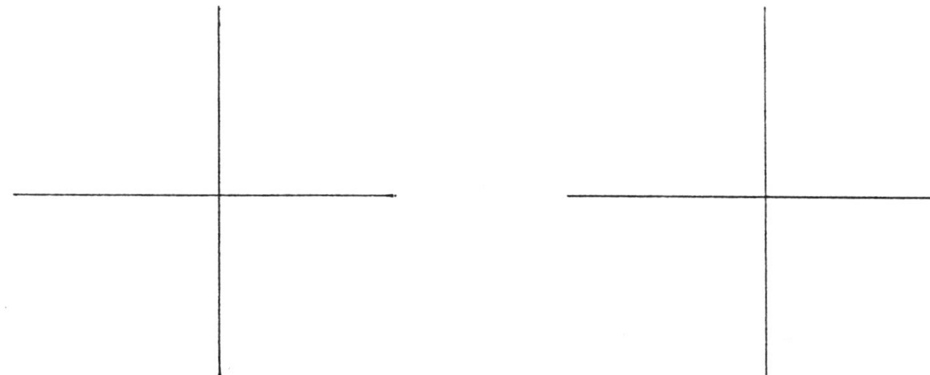

Kinetic Theory - Charles' Law

Laboratory Activity

Charles' law implies that if the volume of a gas is determined at a number of different temperatures, under conditions of constant pressure, the ratio of the volume to the temperature (V/T) will be found to be the same for each. In this experiment, the volume of a sample of air will be measured at a variety of temperatures and the data will be examined to determine if it is consistent with Charles' law.

Materials for each Lab Group:

safety goggles	1 test tube holder
1 250 mL beaker	1 forceps
1 400 mL beaker	1 ring stand
1 glass capillary tube	1 ring support
2mm x 20cm (Can vary)	1 wire gauze
1 watch glass	1 gas burner
1 centimeter ruler	1 spatula
1 thermometer	1 dropper pipet
1 small rubber band	1 ceramic square
dibutyl phthalate	ice
$C_6H_4(CO_2C_4H_9)_2$	water

Procedure:

As you perform the experiment, record your observations in Table 1.

1. Prepare a Charles' law tube. To do this, first hold a glass capillary tube in a test tube holder and seal one end of the tube by rotating it in a hot flame. Then fire polish the open end of the tube.

2. Using a dropper pipet, place 2mL of dibutyl phthalate (DBP) on a clean, dry watch glass. Grasp the prepared capillary tube in a test tube holder. Heat the tube by passing it back and forth through a hot burner flame for approximately 10 seconds. Remove the tube from the heat immediately insert the open end into the DBP on the watch class. Allow DBP to be drawn up into the tube to a height of 1cm.

3. Clamp the capillary tube vertically, with its open end up, on a ring stand, and allow the tube to cool to room temperature. The column of trapped air within the tube should be between 5 and 10cm long.

4. Push a small rubber band or a thin slice of rubber tubing over the capillary tube until it is about halfway along the tube. This rubber band will serve as a distance marker.

5. Fill a 400mL beaker with ice water and position the beaker beneath the clamped capillary tube. Loosen the clamp attachment to the ring stand and lower the tube into the ice water so that the entire column of trapped air is submerged, as shown below.

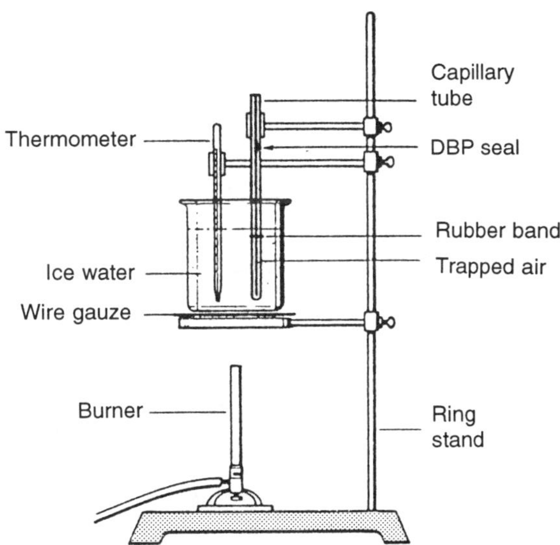

Secure the clamp. Allow the capillary tube and its contents to cool for a few minutes and then, without removing the tube from the ice water, use forceps to carefully push the rubber band along the tube until the upper edge of the band aligns with the top of the air column. Wait another minute and then readjust the band, if necessary.

6. When there is no further change in the height of the air column, measure the temperature of the ice water bath to the nearest 0.1°C. Remove the tube from the bath, being careful to hold it vertically at all times. Measure the distance from the bottom of the tube to the top edge of the rubber band. Record this measurement, and the temperature of the ice water bath, in Table 1.

7. Obtain data for different temperatures by repeating Steps 5 and 6, using water of different temperatures. First, repeat the experiment using a 400mL beaker filled with room temperature water. Then, place the beaker of water on a ring stand assembly and heat the water to 40°C, using a gas burner. Make sure that the water temperature is stable, and not continuing to rise, before you perform the experiment. Also, make sure that for each trial the rubber band has been adjusted properly to mark the top of the air column. Repeat this procedure at temperature intervals of 20°C, up to 100°C, as time permits. Record the height of the air column and the temperature of the bath for each trial.

Calculations:

1. Calculate, in degrees Kelvin, the temperature of the water bath used in each trail. Record the results in Table 1.

2. The volume of gas in each trial is equal to the height of the air column within the capillary tube, multiplied by the cross-sectional area of the tube. However, because the same capillary tube is used in all the trails, the value for its cross-sectional area will be the same (constant) in all the calculations. Therefore, the numerical value for the height of the air column can be used, by itself, to represent the volume in this experiment. Using the measured height of the air column as the volume value, calculate the volume/temperature ratio for each trail and enter your results in Table 1.

3. Enter the ordered pairs (temperature, relative volume) into a statistical data set on your graphing calculator. Using your graphing calculator, draw the scatter plot for this data set and find the best-fitting line (linear regression) for the data points. Write your equation below:

————————————————

4. Replot the scatter plot of the data using degrees Celsius instead of degrees Kelvin. Use Xmin = -300°C and Ymin =0. Again, find the best-fitting line through the data points. Write your equation below:

————————————————

Table 1

Temperature of Water Bath (^{o}C)	Height of Air Column (cm) (relative volume)	Temperature (K)	V/T (cm/K)
________	________	________	________
________	________	________	________
________	________	________	________
________	________	________	________
________	________	________	________
________	________	________	________

Results and Conclusions

1. Is your data consistent with Charles' law? Explain.

2. The pressure did not vary during this experiment because all the trials were performed at constant room pressure. If the pressure had varied, how would it have affected your results? Explain.

3. What factors might contribute to error in this experiment?

Boyle's Law

Content: Inverse relationships, measurement of pressure and volume

Consider the effect of pressure on the volume of a contained gas while the temperature is held constant. When the pressure goes up, the volume goes down. Similarly, when the pressure goes down, the volume goes up. In 1662 the British chemist Robert Boyle proposed a law to describe this behavior of gases. Boyle's Law states that for a given mass of gas at constant temperature, the volume of the gas varies inversely with pressure. The product of pressure and volume of any two sets of conditions is always constant at a given temperature.

The volume of a gas varies as the inverse of the pressure provided the temperature remains constant. Boyle's Law equation is

$$V_1 P_1 = V_2 P_2$$

where:
V_1 = original volume of gas
V_2 = final volume of gas
P_1 = original pressure of gas
P_2 = final pressure of gas

Materials for each Lab Group:

food coloring	screw type pinch clamp
water	small beaker
beral pipet	

Procedure:

1. Fill a small beaker about 1/2 full with water. Add two or three drops of food coloring. Fill the bulb of a beral pipet with the colored water but the stem must remain empty of solution.

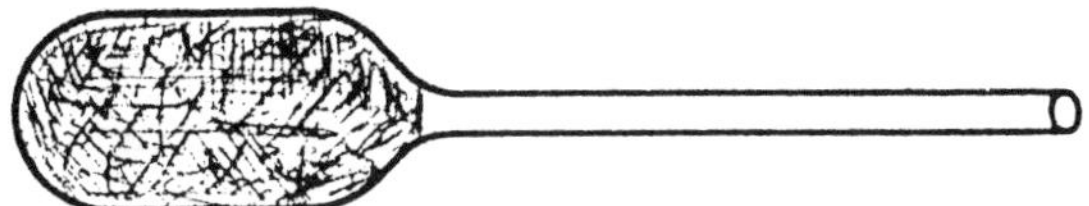

2. Clamp the very end of the beral pipet stem with a screw clamp. The stem now contains a confined gas. What gas is this?

    ____

    The inside diameter of the stem remains constant so the change in volume of the confined gas can be observed. As pressure is applied to the bulb of the pipet, some of the colored water will move into the stem of the pipet. The volume of confined gas can be measured in the stem of the pipet.

3. Place the beral pipet on the lab table. Using a centimeter ruler measure the length of the gas sample in the stem of the beral pipet. This in the initial volume of the confined gas. Place a book on the bulb of the pipet. Pressure is equal to the number of books and volume is equal to the length of the gas trapped in the pipet stem. Record all pressure and volume measurements in Table 1.

4. Place another book on top of the first and record the new pressure and volume. Try to keep the books in the same position on top of the beral pipet. Continue increasing the pressure (books) on the confined gas. The limit of pressure (books) is 10.

5. If time permits, test the beral pipet with "infinite pressure". Place the beral pipet on the floor with one book on top as before. One person will supply "infinite pressure" by carefully standing on the book.

Table 1

PRESSURE (books)	VOLUME (mm)	P x V
0	__________	__________
1	__________	__________
2	__________	__________
3	__________	__________
4	__________	__________
5	__________	__________
6	__________	__________
7	__________	__________
8	__________	__________
9	__________	__________
10	__________	__________

Results and Conclusions:

1. Enter the ordered pairs (pressure, volume) into a statistical data set within your graphing calculator (leave out the pressure of zero books). Draw the scatter plot for the data in a viewing window of Xmin = 0, Xmax = 11, Xscl = 1, Ymin = 0, Ymax = 200, and Yscl = 25. Pressure in "books" is plotted on the X-axis and volume in 'mm' is plotted on the Y-axis. Use your graphing calculator to find the best-fitting line for the data (linear regression). Write the equation below:

    ____

    Graph this equation on your graphing calculator in the viewing window given above. Use the trace feature to approximate the volume at a pressure of 6.5 books.

    ____

2. Looking at the scatter plot and the slope of the equation above, determine the relationship between the pressure and the volume of the gas?

    ____

3. Using the information from Table 1, calculate the product of pressure times volume for each trial. Record each product in the data table. What is noticeable about the product of pressure x volume?

    ____

4. What would happen to the gas at very high pressures? (Assume the "infinite pressure" did not cause the beral pipet to split.)

    ____

Introduction to the Spectrophotometer

Laboratory Activity

Content: Absorption spectrum, instrumentation and graphing

Using a spectrophotometer, we will obtain the spectrum of a known ion in aqueous solution. A spectrophotometer is made up of a radiation source, a diffraction grating, a monochromator, and a detector. The radiation source will emit radiation at all wavelengths in the visible and ultraviolet regions of the electromagnetic spectrum. This continuous radiation is dispersed into a spectrum by means of the diffraction grating. The purpose of the monochromator is to select a narrow band of wavelengths, which will be directed to the detector. The detector will measure the intensity of the wavelengths of light transmitted from the monochromator.

A sample will be placed in the light path between the monochromator and the detector. The sample may then transmit all of the light, some of the light or no light. This will depend on the nature of the material in the sample. Light absorption occurs at wavelengths where the energy of the light corresponds to the energy necessary to cause electronic excitations of atoms, ions, or molecules in the sample. These absorptions will appear as a depressions on a graph using absorption versus wavelength. This type of graph is called an absorption spectrum.

Why are absorption spectra useful to us? First, because the absorption spectrum of a particular substance is characteristic of that substance. This is helpful to us for identification purposes. We know that our fingerprints are characteristic to us as individuals. This is also true of the absorption spectrum. Second, the intensity of the absorption bands can be related to the concentration of the substance in the sample. Therefore, it is possible to determine the amount of a particular substance in a mixture.

The absorption spectrum of an aqueous solution of chromium (III) ions will be determined in this experiment.

Materials for each Lab Group:

Spectrophotometer	10 mL graduated cylinder
small test tubes or cuvettes	safety goggles
plastic wash bottle	tissue paper
distilled water	0.02M chromium (III) nitrate
	$Cr(NO_3)_3$

CAUTION!! Chromium (III) nitrate is toxic and can irritate your skin. Avoid contact with this chemical. Wash your hands thoroughly after use. Do not pick up dropping bottles by their tops.

Procedure:

1. A warm-up of 20 minutes is usually required for the spectrophotometer. Your teacher will have already turned the instrument on and adjusted the amplifier control knob to read zero percent transmittance (0%T). After the instrument has warmed up, adjust the needle back to zero if necessary.

2. Set the wavelength dial at 375 nanometers (375 nm) using the wavelength control knob. Add Adjust the amplifier control knob to 0%T at this wavelength. The cuvette cover must be kept closed when adjusting the control knob and making readings.

3. Add 3 mL of water to a clean small test tube or cuvette. This tube is called a "blank" because it does not contain any of the substance being tested. (NOTE: It is very important that the tube be dry on the outside and that it also be clean. Dirt on the test tube or water drops on the outside of the tube can affect light transmittance and can be the source of erroneous results.) Holding the "blank" at its top, wipe the outside of the tube with a tissue to make certain that it is clean and dry.

   Do not get finger prints on the outside of the sample tube. Air bubbles clinging to the inside walls of the sample tube can also affect light transmittance. Any bubbles should be dislodged by gently tapping the side of the tube before any readings are taken.

Place the tube "blank" in the sample holder of the spectrophotometer. Close the cover of the holder. Adjust the light control knob until the percent transmittance (%T) reads 100.

4. Remove the "blank" from the spectrophotometer and save it for use later. Add 3 mL of 0.02M chromium (III) nitrate to a second test tube or cuvette. Wipe the tube clean and dry with a tissue and insert the tube into the sample holder. Close the cover of the holder. Read and record the percent transmittance (%T) from the meter in Table 1. Remove the sample from the holder.

5. Turn the wavelength dial to 400 nm. Adjust the percent transmittance (%T) to zero, using the amplification control knob. Place the "blank" in the holder, close the cover, and adjust the meter to 100%T with the light control knob. Put the 0.02 M $Cr(NO_3)_3$ solution in the holder again and read the %T at 400 nanometers and record the value in Table 1.

6. Continue the procedure of setting 0%T, setting 100%T (with blank in spectrophotometer), place tube with Cr(III) ions in spectrophotometer, and measuring the %T of the chromium (III) nitrate solution at 405, 415, 425, 440, 455, 470, 490, 500, 530, 540, 550, 570, 575, 580, 600, and 625 nm.

7. Return the aqueous chromium (III) nitrate solution to the waste container.

8. Clean and return all equipment and chemicals to proper location in lab.

Table 1:

Percent Transmittance of 0.02M $Cr(NO_3)_3$ Solution

Wavelength (nm)	% Transmittance (%T)
375	__________
400	__________
405	__________
415	__________
425	__________
440	__________
455	__________
470	__________
490	__________
500	__________
520	__________
530	__________
540	__________
550	__________
570	__________
575	__________
580	__________
600	__________
625	__________

Data Analysis:

1. Enter the ordered pairs (wavelength, % transmittance) into a statistical
    data set within your graphing calculator. Draw the scatter plot within
    a viewing window of Xmin = 350, Xmax = 650, Xscl = 25, Ymin = 0,
    Ymax = 100, and Yscl = 10. The wavelength is represented on the X
    axis and the % transmittance on the Y axis.

2. The data points in the scatter plot form a curve of the absorption
    spectrum of Cr(III) ion in the visible region. Sketch this graph in the
    box provided below:

Results and Conclusions:

1. What is the purpose of the "blank" used in the lab?

2. At what wavelengths do Cr(III) ions absorb the maximum amounts
    of light?To what colors of light do these wavelengths correspond?

3. When a solution is red, does it absorb or transmit red light? Explain.

4. Using accepted percent transmittance (%T) values provided
    by your teacher, calculate % error for 400 nm, 470 nm, and
    575 nm wavelength measurements.

Freezing Point Depression Analysis of the Ice Cream Process

Laboratory Activity

Content: temperature depression due to solvent

When an ionic substance is added to a solvent we generally see a drop or depression in freezing point. This can be traced graphically and used to show different effects generated by different variables. In this lab we will make and interpret several graphs of different variables.

Materials for each Lab Group:

Thermometer	Beaker
Ice Cream Freezer	Bunsen Burner
Ice Cream Mix	Wire Gauze
Salt	Ring Stand
Ice	Igniter

Procedure:

1. Fill the beaker to the top with ice and level the ice to the lip of the beaker. Melt the ice and find the volume of water that is held in the ice by pouring the melt water into a graduated cylinder. Record your finding.

2. Assemble your ice cream freezer and your teacher will fill the freezer with the required amount of mix.

3. Pack ice around your ice cream freezer. Make sure each beaker of ice is leveled so your measurements will be accurate. Record the number of beakers of ice that you add. You will add ice three times, so be sure and put these readings in order.

    a.__________   d.__________
    b.__________   e.__________
    c.__________   f.__________

4. Measure the temperature of the ice with your thermometer. Do not measure the temperature with the motor running. Record the temperature in Table 1.

5. Without the motor running add 58 g. of NaCl to the ice.

6. Measure the temperature and record it in Table 1.

7. Run the motor for two minutes then measure the temperature.

8. Continue to measure the temperature until the temperature drop stops. Record each temperature.

9. Once a plateau is reached add another 58 g. of NaCl and enough ice to fill the ice bucket. It is important that you record the amount of ice and salt added.

10. Follow steps 7 through 10 until the ice cream freezer stops. Do not open the freezer cylinder until the motor completely stops otherwise you may ruin the contents with salt contamination.

11. Following the lab please "properly" dispose of the contents of the freezer cylinder.

Analysis:

1. Enter the ordered pairs (time, temperature) from Table 1 into a statistical data set within your graphing calculator. Draw the scatter plot of the data points in a viewing window of Xmin = 0, Xmax = 30, Xscl = 2, Ymin = -15, Ymax = 1, and Yscl = 2. Sketch your scatter plot in the box provided below:

2. Enter the ordered pairs (concentration, temperature drop) from Table 1 into a statistical data set within your graphing calculator. Use the final temperature drop for each concentration, for example if the temperature drops 4° with 58 g of NaCl this would be the drop due to this amount of salt.

Next, figure your molality and use it for the concentration. Draw
the scatter plot of the data points in anappropriate viewing
window. Sketch your scatter plot in the box provided below:

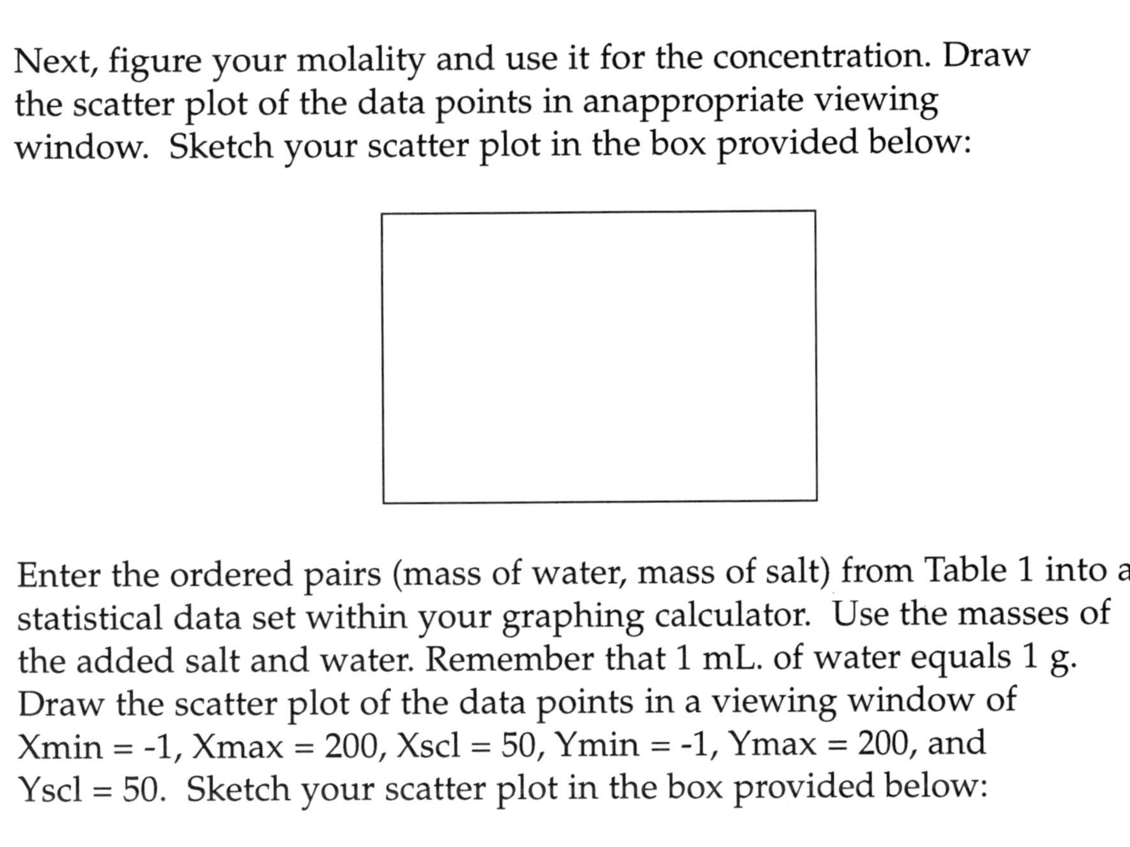

3. Enter the ordered pairs (mass of water, mass of salt) from Table 1 into a
    statistical data set within your graphing calculator. Use the masses of
    the added salt and water. Remember that 1 mL. of water equals 1 g.
    Draw the scatter plot of the data points in a viewing window of
    Xmin = -1, Xmax = 200, Xscl = 50, Ymin = -1, Ymax = 200, and
    Yscl = 50. Sketch your scatter plot in the box provided below:

Table 1:

Time	Temperature	Mass of Salt	Volume of H_2O
1.	__________	__________	__________
2.	__________	__________	__________
3.	__________	__________	__________
4.	__________	__________	__________
5.	__________	__________	__________
6.	__________	__________	__________
7.	__________	__________	__________
8.	__________	__________	__________
9.	__________	__________	__________
10.	__________	__________	__________
11.	__________	__________	__________
12.	__________	__________	__________
13.	__________	__________	__________
14.	__________	__________	__________
15.	__________	__________	__________
16.	__________	__________	__________
17.	__________	__________	__________
18.	__________	__________	__________
19.	__________	__________	__________
20.	__________	__________	__________

Exercises:

1. From your scatter plots what type of relationships do we see between the temperature drop and our concentration? What mathematical equation could you derive from this relationship?

____

____

____

2. Using the (mass of water, mass of salt) data, determine the molality of the solution at each data point. What differences in molality did you expect? What were the actual changes?

____

____

____

3. From the scatter plots you have seen, why is the (mass of water, mass of salt) plot different from the other plots?

____

____

____

4. Why did we see the temperature stabilize?

____

____

____

Solubility

Laboratory Activity

Content: Solubility and graphing

If you have ever opened a hot can of soda pop you have been zapped by a factor of solubility. We have learned over the years that several different factors can and do effect the solubility of certain substances. Since we use solutions on a daily basis we have had to learn how to manipulate these factors to our advantage.

In this lab we will explore several different factors in solubility. It is very important that you pay close attention to detail or you may miss a reading. We will use graphs to help show the affects of different factors on our rate of solution. You will need to write all your data down in Table 1.

Materials for each Lab Group:

6 - large test tube	Rock salt	KNO_3
3 - 250 mL. beaker	Table salt	KCl
Hot plate*	Water	$NaNO_3$
Thermometer	Ice	Sugar
Graduated Cylinder	Stirring Rods	Filter paper
Centigram balance	Beaker Tongs	Stopwatch

*If hot plate is not available you will need to set up a lab burner and a ring stand to heat the beakers of water.

Procedure:

1. Using the graduated cylinder measure out 150 mL. of water and pour into a beaker. Place 150 mL. of water in each of the beakers.

2. Place one of the beakers on the heating apparatus and begin heating it to a temperature of 85°C. Once the water reaches the desired temperature remove it from the heat. Don't let the heat get below 80°C.

Do not pick up the hot beaker, use the beaker tongs.

3. Place ice in one of the other beakers. Before you use the beaker check the temperature and make sure it is below $10\,^\circ C$.

4. Place 5 mL. of water in three of the test tubes after you have measured out the water in the graduated cylinder.

5. Place one of the test tubes in each of the three different beakers of water. Check the temperature in each beaker.

6. Add .25 grams of salt to each of the three test tubes. Begin timing immediately after you add the salt to the test tube. Do not stir the solution. Enter time in Table 1.

7. Stop the time once the salt is completely dissolved. Record the time in Table 1.

8. Dispose of the solutions as your teacher instructs and clean the test tubes for the next procedure.

Procedure:

1. Label four clean test tubes KNO_3, KCl, $NaNO_3$, NaCl. Place 5 mL of water in each test tube.

2. Check the warm water bath and make sure the temperature is in $85\,^\circ C$ range.

3. Add 2 g of the chemical that is written on the label of the test tube. Place the test tube in the warm bath. Immediately start the timing. Record value in Table 2.

4. Stop the stopwatch when the chemical is totally dissolved. Record the data in Table 2.

5. Repeat steps 3 and 4 for each chemical.

6. Dispose of the solutions as your teacher instructs.

7. Add 5 mL of water to the test tubes.

8. Check the cold water bath to make sure it's below 10 °C.

9. Add 2 g of the chemical that is written on the label of the test tube. Place the tube in the cold water bath and start the time. Record value in Table 2.

10. Keep time until the chemical is dissolved. Record the time in Table 2.

11. Repeat steps 8-10 for each chemical and dispose of the chemical as you did in step 6.

12. Add 5 mL of water to each of the test tubes.

13. Place 2 g of the chemical that is written on the label, into the test tube.

14. Place the test tube into the beaker of water at room temperature. Start keeping time. Record value in Table 2.

15. Stop the time once all the chemical is dissolved. Record value in Table 2.

16. Repeat step 13-15 for each chemical.

17. Dispose of the solutions and clean your test tubes.

Procedure:

1. Place 5 mL. of water in two clean test tubes.

2. Add 2.0 grams of rock salt to one test tube and 2.0 grams of granulated salt to the other test tube.

3. Time how long it takes the two different types of salt to dissolve completely. Record the times in Table 3.

4. Clean your test tubes and dispose of the solution as you did earlier.

Procedure

1. Place 5 mL. of water in two clean test tubes.

2. Add 2.0 grams of granulated KCl to each test tube.

3. Begin the time once the KCl is added to the water. Record time in
    Table 4.

4. Choose one of the test tubes to stir and the other one to sit.

5. Stir the chosen test tube as rapidly as possible. Do not disturb the
    other test tube.

6. Stop the time once the KCl is completely dissolved in both tubes.
    Be sure and record the amount of time necessary for the salt to
    dissolve in each test tube. Record the times in Table 4.

7. Dispose of the solution and clean the test tubes.

1. Enter the three ordered pairs (temperature, elapsed time) from Table 1 into a statistical data set within your graphing calculator. Draw a scatter plot for the data on the graphing calculator within an appropriate viewing window. Sketch this scatter plot in the box provided below:

Use your graphing calculator to find the best-fitting line for the data points (linear regression). Write this equation below:

____

Graph this equation in the same viewing window as the scatter plot. Use the tracer to predict the time necessary at $30^{\circ}C$.

____

2. Enter the three ordered pairs for KNO_3 (temperature, elapsed time) from Table 2 into a statistical data set within your graphing calculator. Draw a scatter plot for the data points on the graphing calculator within an appropriate viewing window. Sketch this scatter plot in the box provided below:

Use your graphing calculator to find the best-fitting line for the data
points (linear regression). Write this equation below:

____

Graph this equation in the same viewing window as the scatter plot.
 Use the tracer to predict the time necessary at 30°C.

____

3. Repeat #2 for KCl, NaNO$_3$, and NaCl.

KCl

Equation: _________________   Time: _______________

NaNO$_3$

Equation: _________________   Time: _______________

NaCl

Equation: _______________   Time: _______________

Questions:

Answer each question completely. Use the data table, scatter plots, best-fitting lines, and your lab notes to answer the questions.

1. Describe the solubility of each of the chemicals that you graphed in second procedure. Use descriptions like most soluble, 2nd soluble, 3rd soluble, and least soluble.

	Cold Water	Room Temp. Water	Hot Water
KNO_3	_________	_________	_________
KCl	_________	_________	_________
NaCl	_________	_________	_________
$NaNO_3$	_________	_________	_________

2. How does temperature effect solubility in some chemicals?

____

____

3. What differences did you see in the granulated salt and the rock salt when we tried to dissolve them?

____

____

4. How did stirring effect the rate of solution in the KCl?

____

____

5. From this lab, what would you describe as the major factors in causing a solid to go into solution with a liquid? Will your methods always work? Why or Why not?

____

____

____

____

Table 1

	Temp.	Beginning Time	Ending Time	Elapsed Time
Ice Water	_______	_______	_______	_______
Room Temp. Water	_______	_______	_______	_______
Hot Water Bath	_______	_______	_______	_______

Table 2

	Cold Water Temp=_______			Room Temp. Temp=_______			Hot Water Temp=_______		
	Start	End	Elap	Start	End	Elap	Start	End	Elap
KNO_3	___	___	___	___	___	___	___	___	___
KCl	___	___	___	___	___	___	___	___	___
$NaNO_3$	___	___	___	___	___	___	___	___	___
NaCl	___	___	___	___	___	___	___	___	___

Table 3

	Beginning Time	End Time	Elapsed Time
Rock Salt	_____________	_____________	_____________
Granulated Salt	_____________	_____________	_____________

Table 4

	Beginning Time	End Time	Elapsed Time
Stirred KCl	_____________	_____________	_____________
Calm KCl	_____________	_____________	_____________

Boiling Point Elevation

Laboratory Activity

Content: Temperature elevation

When a solute is added to water you will generally see a change in the boiling point. The change is due to the action of the solute in the solution. When the solute dissolves or dissociates the molecules are better able to absorb heat. Since ionic compounds dissociate and give off more than one hydrated particle you see that they tend to affect the boiling point more than covalent compounds.

In this lab, you will use three different chemicals in two different concentrations. You are expected to graph the changes in the boiling point due to the chemicals added.

Materials for each Lab Group:

Lab burner	Graduated cylinder	igniter
3 - 250 mL. beakers	Thermometer	$CaCl_2$
ring stand	iron ring	NaCl
hot pad	asbestos pad	Table sugar

Procedure:

1. Set up the boiling apparatus. Make sure the iron ring fits the beakers.

2. Measure out 150 mL. of distilled water in a graduated cylinder.

3. Place the distilled water in the 250 mL beaker and place the beaker on the ring stand.

4. Heat the water until it begins to boil.   Measure and record the temperature. This will standardize your thermometer.

5. Obtain 150 mL. of a .10 M solution of the table sugar.

6. Boil this solution and determine the boiling point using the same

procedure as in step 4. This procedure should be followed for the
.10 M solutions of calcium chloride and sodium chloride. Record the
boiling points for all of these solutions in Table 1.

7. Rinse out all the beakers after they have cooled sufficiently to handle.
Be sure that no residue is left inside the beaker.

8. Obtain 150 mL. of the .40 M solution of table sugar. Once again, record
the temperature in Table 1 when the solution begins to boil.

9. Obtain 150 mL. of the .40 M solution of the calcium chloride and the
sodium chloride. Follow the same procedure for these solutions.
Record the boiling point for the .40 M solutions in Table 1.

10. Dispose of the chemicals properly and clean the lab area.

Analysis:

1. Enter the ordered pairs (solution number, temperature) for each of the
.10 M solutions in Table 1. The solution number for distilled water is 1,
whereas table sugar is 2, sodium chloride is 3, and calcium chloride is
4. Using your graphing calculator, draw the line graph (a line is drawn
from one point to the next in a connect-the-dot fashion) for the data
points in a viewing window of Xmin = 0, Xmax = 5, Xscl = 1,
Ymin = 99, Ymax =103, and Yscl = 1. Sketch your line graph below:

2. Repeat #1 for the .40 M solutions.

Questions:

1. By using their chemical formulas predict which chemical we tested
   will be most like the following chemicals:
   a) Lithium fluoride   ___________________
   b) Beryllium chloride   ___________________
   c) Potassium bromide   ___________________
   d) Magnesium iodide   ___________________

2. Why did the table sugar not cause as much elevation as the other
   chemicals?

   ____

   ____

   ____

   ____

3. Using your line graphs, what would you expect to happen if we
   increased the concentration of the table sugar, salt and calcium
   chloride in the solution?

   ____

   ____

   ____

   ____

Table 1: Boiling Point Temperature

	.10 M Solution	.40 M Solution
Distilled Water	______________	______________
Table Sugar	______________	______________
Sodium Chloride	______________	______________
Calcium Chloride	______________	______________

The Dreaded M's

Pre-Laboratory Activity

Content: Differentiating between molarity and molality, compute molality, and graph molality

The use of these two terms is essential to describing the concentration and effect that different solutions or solutes will have in a solution. We will generally try to use the molarity when we discuss acids and bases. Molality is very useful in describing how much solute was used to mix the base chemical. Our purpose in this lab is to make you use both molarity and molality to find the mass, volume and desired type of concentration from the original concentration type you were given.

The important thing to remember is that molality and molarity change when we change the amount of solute or solvent. The other important thing to remember is that molality and molarity are not the same term. They are derived differently and show different things. They can be used to derive each other if the right information is present.

Solving for molality from mass:

Solving for molarity from mass:

$$\frac{\text{Mass of solute}}{\text{Mass solvent}} \times \frac{1 \text{ mol}}{\text{mass of } 1 \text{ mole solute}} \times \frac{1000 \text{ g}}{\text{kg}} \qquad \frac{\text{Moles of solute}}{\text{Liters of solvent}}$$

Practice finding the molarity. Assume all solvents are H_2O.

a. 18 g of HCl in .25 L   __________

b. 36 g of HCl in 2.5 L   __________

c. 72 g of HCl in 2 L   __________

Enter the ordered pairs (moles, volume) from preceding three problems into a statistical data set within your graphing calculator. Draw a scatter plot for the data within an appropriate viewing window (limits should include all your mole and volume values). Sketch your scatter plot in the box provided below:

Using the three problems from above:

1. Find the molality of each mass of HCl. Remember that 1 mL of water is equal to 1 gram of water. It takes 1000 g. to make a liter.
   a. _____________
   b. _____________
   c. _____________

2. Enter the ordered pairs (moles, volume) from preceding three problems(use data from #1 above) into a statistical data set within your graphing calculator. Draw a scatter plot for the data within an appropriate viewing window (limits should include all your mole and volume values). Sketch your scatter plot in the box provided below:

   Do the shapes of the scatter plots for molarity and molality change?

Reaction Rate

Laboratory Activity

Content: Data on concentration vs. time and on temperature vs.
                time; and scatter plots of data.

In this experiment, two solutions will be mixed, and the completion of the
reaction will be marked by a color change. One solution contains the
iodate ion(IO_3^-). The other contains the hydrogen sulfite ion(HSO_3^-) and
soluble starch. In the presence of starch molecules, molecular iodine(I_2)
produces a characteristic blue color. The rate of the entire reaction can be
determined by timing the interval between the time the two solutions are
mixed and the appearance of the blue color. By varying the concentration
of one of the reactants (at constant temperature) and then varying the
temperature alone, you can observe and record the effects of these two
factors on reaction rate.

*Note: Half the class can do Part 1 and half Part 2 or this can be done over a
two day period.*

Materials for each Lab Group

    beaker-250 mL   2 beakers-100 mL
    2 graduated cylinders-10 mL   2 test tubes-18x150 mm
    thermometer   Lab apron or coat
    Safety goggles   Timer (stop watch or
    Distilled Water   clock with second hand)
    Ice cubes

Solution A: Dissolve 4.3g of potassium iodate (KIO_3) per liter
            of solution to produce a 0.02M solution

Solution B: Dissolve 2g of sodium bisulfite ($NaHSO_3$),5 mL of
            1 M H_2SO_4, and 4g of soluble starch per liter of
            solution

*Note: For Part 1 each Lab Group will need Sol.A-150mL, Sol.B-105mL, and
distilled water-45mL.*

Procedure: Part 1

1. Using a clean, dry, 10mL graduated cylinder, measure exactly 15.0mL
    of Solution A and pour it into a 100ml beaker.

2. Using a second 10mL graduate, measure exactly 15.0ml of Solution B
    and pour it into a second 100ml beaker.

3. Prepare to time the reaction. While one lab partner pours Solution A
    into Solution B, the second partner should immediately start timing the
    reaction. Pour the solutions back and forth several times from one
    beaker to the other to ensure thorough mixing. Then allow the mixture
    to stand. At the instant a color change occurs, the partner timing the
    reaction should note the elapsed time. Record this in Part 1 of Table 1.
    Rinse and dry the beakers and graduated cylinders.

4. Measure exactly 15.0ml of Solution B into one of the beakers. Measure
    exactly 14.0ml of Solution A into the other beaker. Dilute this solution
    by adding exactly 1.0 mL of distilled water. Follow the instructions in
    step 3 for mixing the solutions and timing the reaction. Record the
    elapsed time in your data list. Rinse and dry the beakers and
    graduated cylinders.

5. Repeat step 4 eight more times, using increasingly dilute samples of
    Solution A. Use the following ratios of Solution A to distilled water
    (in mL): 13 to 2; 12 to 3; 11 to 4; 10 to 5; 9 to 6; 8 to 7; 7 to 8; and 6 to 9.
    Rinse and dry beakers and graduated cylinders after each trial. Record
    all times in the data list.

Note: For Part 2 each Lab Group will need Sol. A-50mL, Sol. B-50mL, and ice.

Procedure: Part 2

1. Measure 10.0mL of Solution A into one test tube and 10.0mL of Solution B into a second test tube.

2. Half fill a 250mL beaker with cold tap water. Add ice cubes to the water and stir carefully with thermometer. Continue stirring (and adding ice as needed) until the temperature of the ice-water mixture is about 5°C.

3. Place the two test tubes in the ice-water bath and let them stand until the solutions are at the same temperature as the ice water. (Always rinse and wipe the thermometer after removing it from a solution.)

4. When the solutions are at the same temperature as the ice water, prepare to time the reaction. One lab partner should start timing the reaction the instant the second partner pours Solution A into Solution B. Quickly pour the mixture back and forth from test tube to test tube several times and return the mixture to the ice-water bath. At the instant a color change occurs, note the time elapsed. Measure the temperature of the mixture immediately. Record the exact temperature and elapsed time in your data table. Discard the mixtures as instructed. Rinse and dry the test tubes.

5. Repeat step 1.

6. Prepare a water bath at a temperature of about 10°C. Repeat steps 3 and 4 at this new temperature. Record your observations in Table 1.

7. Repeat these procedures using warm baths at the following temperatures: 15°C; 20°C; 25°C; 30°C; 35°C; 40°C. Use warm tap water to prepare these baths. Rinse and dry the test tubes after each trial.

Table 1

PART 1

Solution B (mL)	Solution A (ml)	H_2O (mL)	Time(sec)
15	15	0	__________
15	14	1	__________
15	13	2	__________
15	12	3	__________
15	11	4	__________
15	10	5	__________
15	9	6	__________
15	8	7	__________
15	7	8	__________
15	6	9	__________

PART 2

Trial #	Temperature(^{o}C)	Time(sec)
1	5°	__________
2	10°	__________
3	15°	__________
4	20°	__________
5	25°	__________
6	30°	__________
7	35°	__________
8	40°	__________

Calculations:

1. Enter the ordered pairs (mL H_2O, time) from Part 1 of Table 1 into a statistical data set within your graphing calculator. Using the graphing calculator, draw the scatter plot for the data within a viewing window of Xmin = -1, Xmax = 10, Xscl = 1, and Ymin = 0. Set the Ymax and Yscl appropriately for the data observed. Draw the scatter plot in the box provided below:

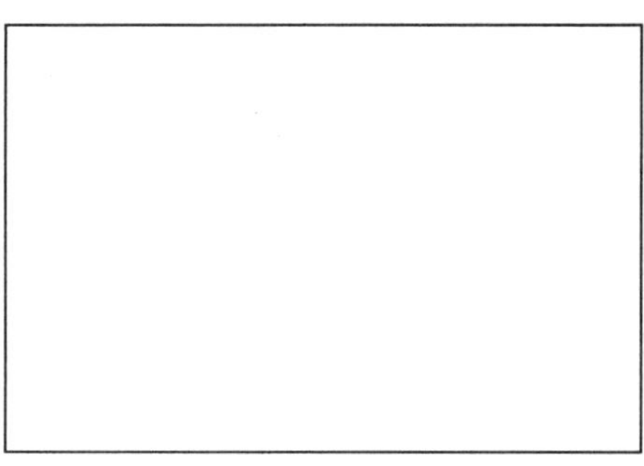

Use your graphing calculator to find the best-fitting line for the data points (linear regression). Write this equation below:

____

Graph this equation and use the tracer to predict the time necessary at 5.5 mL of H_2O.

____

2. Enter the ordered pairs (temperature, time) from Part 2 of Table 1 into a statistical data set within your graphing calculator. Using the graphing calculator, draw the scatter plot for the data within a viewing window of Xmin = 0, Xmax = 45, Xscl = 5, and Ymin = 0. Set the Ymax and Yscl appropriately for the data observed. Draw the scatter plot in the box provided:

Use your graphing calculator to find the best-fitting line for the data points (linear regression). Write this equation below:

____

Graph this equation and use the tracer to predict the time necessary at 22°C.

____

Heat of Fusion and Vaporization

Pre-Laboratory Activity

Content: Melting point, boiling point, heat of fusion, and heat of vaporization.

If we use q to represent the quantity of energy needed to melt a substance then $q=m(H_{fus})$ where m is the mass of the substance and H_{fus} is a property of a substance (Heat of Fusion). Similarly, to boil a substance, the relationship is $q=m(H_{vap})$ where H_{vap} is a property of a substance (Heat of Vaporization).

The amount of energy required to change the temperature of a substance depends upon the amount and nature of the substance as well as the temperature change. The energy required is calculated by $q=m(\Delta T)C_p$ where q is the energy added or removed and m is the mass of the substance, ΔT is the change in temperature, and C_p is a property of the substance called specific heat.

You may be asked to do calculations that involve both temperature and state changes. For such problems, each step involving a temperature or state is solved separately. The sum of the energy changes for all the steps is the solution to the problem.

For example, suppose you are asked to start with a solid substance below the melting point and determine the energy required to change that substance to a gas above its boiling point. The following steps must be taken:

a. Heat the solid to its melting point.
b. Melt the solid.
c. Heat the liquid to the boiling point.
d. Boil liquid.
e. Heat gas to required temperature.
f. Add all results.

1. Calculate the heat needed raise the temperature of
    5.58 Kg of iron from 20° C to 1000° C.
    (C_p of iron is .448 J/g*°C)

2. Use your graphing calculator to graph the equation for the heat needed
    to raise the temperature of 5.58 Kg of iron from 20°C. The equation to
    be graphed is q = (5580)(ΔT)(.448). This can be entered into your
    graphing calculator as Y1 = 5580*.448*X. Graph the equation in a
    viewing window of Xmin = 0, Xmax = 1000, Xscl = 100,
    Ymin =0, Ymax = 3000, Yscl = 500. Sketch the graph in the box
    provided below:

3. Use the tracer to find the amount of heat needed to raise the
    temperature of 5.58 Kg of iron from 20°C to 900°C.
    Remember, you are looking for an X of 880 while tracing.

    ____

4. Calculate the amount of energy necessary to convert 10 g of water at 10° C to steam at 150° C.

(H$_{vap}$ of water 2260 J/g.)
(C$_p$ of water 4.18 J/g*$^\circ$C)
(C$_p$ of steam 2.02 J/g*$^\circ$C)

5. Calculate the amount of energy necessary to convert 42.5 g of aluminum vapor 4750° C to 25° C.

(Freezing point of Al is 660° C)   (H$_{vap}$ = 291 KJ/mol)
(Boiling point of Al is 2467° C)   (H$_{fus}$ = 10.5 KJ/mol)
(C$_p$(gas)=1.05 J/g*$^\circ$C)   (C$_p$(l)=.869 J/g*$^\circ$C)
(C$_p$(cr)=.900 J/g*$^\circ$C)

(Hint: q=m*mole conversion*C$_p$)